World of Robots

I0894015

By Elizabeth Noll

Military Robots

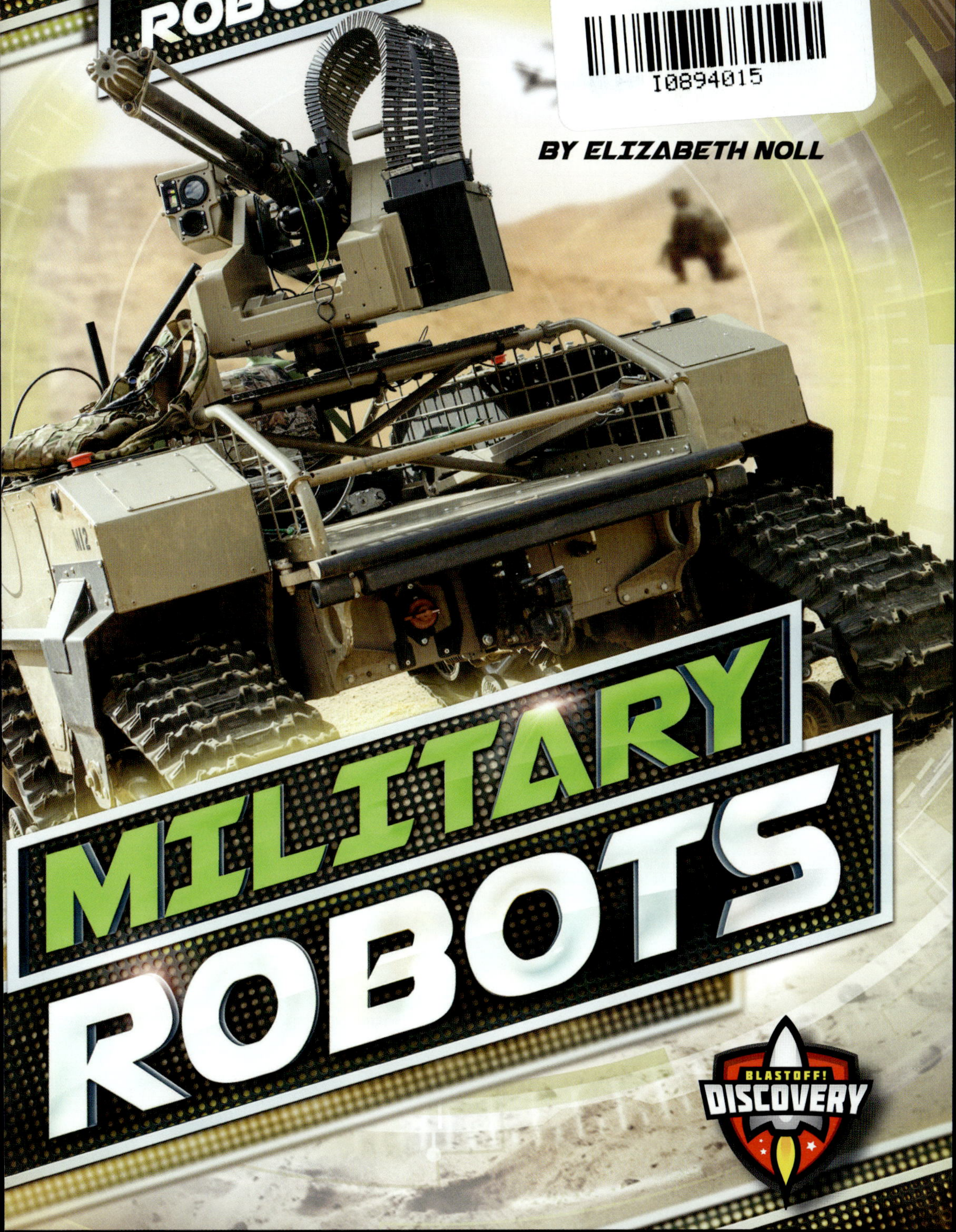

Blastoff! Discovery

Bellwether Media • Minneapolis, MN

Blastoff! Discovery launches a new mission: reading to learn. Filled with facts and features, each book offers you an exciting new world to explore!

This edition first published in 2018 by Bellwether Media, Inc.

No part of this publication may be reproduced in whole or in part without written permission of the publisher.
For information regarding permission, write to Bellwether Media, Inc., Attention: Permissions Department, 5357 Penn Avenue South, Minneapolis, MN 55419.

Library of Congress Cataloging-in-Publication Data

Names: Noll, Elizabeth, author.
Title: Military Robots / by Elizabeth Noll.
Description: Minneapolis, MN : Bellwether Media, Inc., [2018] | Series: Blastoff! Discovery. World of Robots | Audience: Ages 7-13. | Includes bibliographical references and index.
Identifiers: LCCN 2016057454 (print) | LCCN 2016058097 (ebook) | ISBN 9781626176904 (hardcover : alk. paper) | ISBN 9781681034201 (ebook) | ISBN 9781618912930 (paperback : alk. paper)
Subjects: LCSH: Military robots–United States–Juvenile literature.
Classification: LCC UG450 .N65 2018 (print) | LCC UG450 (ebook) | DDC 623–dc23
LC record available at https://lccn.loc.gov/2016057454

Text copyright © 2018 by Bellwether Media, Inc. BLASTOFF! READERS and associated logos are trademarks and/or registered trademarks of Bellwether Media, Inc. SCHOLASTIC, CHILDREN'S PRESS, and associated logos are trademarks and/or registered trademarks of Scholastic Inc., 557 Broadway, New York, NY 10012.

Editor: Christina Leaf Designer: Jon Eppard

Printed in the United States of America, North Mankato, MN.

TABLE OF CONTENTS

MILITARY ROBOT AT WORK!

Mohammed stood by the edge of the road. Around him, other soldiers were quiet. They needed to get back to **base camp** before nightfall. But a villager had told them that the enemies buried a bomb nearby.

The soldiers lifted a metal object from the back of their truck. Then, they took cover behind the truck. Mohammed opened a hard case and pressed a button. He watched on a screen as TALON, a small squat robot, rolled forward.

Suddenly the road ahead jumped. Fire and smoke shot upward. The robot flew into the air. Shock waves slammed the soldiers into the ground. For a long time, Mohammed lay still. His ears were ringing. His head hurt.

When the soldiers got up, they saw a huge crater in the road. If they had gone a little farther, they would have been blown up. Mohammed pressed another button. The robot righted itself and started toward him. TALON had saved the soldiers' lives.

WHAT ARE MILITARY ROBOTS?

Many countries use robots for their militaries, including the United States. In the military, the term for robots is unmanned ground vehicle (UGV) or unmanned aerial vehicle (UAV). However, many people know UAVs as drones.

Ground robots can be "man-portable," meaning small enough for a person to carry. Others are as big as a truck. There are many different sizes of UAVs, too. Pocket-sized drones can be used for **surveillance**. Some drones are large enough to carry a Hellfire **missile**.

man-portable
UAV

Hellfire missiles
MQ-9 Reaper UAV

Multi-Utility Tactical
Transport Vehicle
(MUTT)

The most important thing military robots do is save lives. Many military robots travel ahead of soldiers to test the ground. Without these machines, soldiers are in danger of stepping on **land mines** and other explosives. Robots keep soldiers safe from this scary part of war.

Military robots save lives in other ways, too. Some carry supplies that soldiers need to survive. Others can go into places too dangerous for humans. Toxic chemicals and **nuclear radiation** do not harm robots.

soldiers inspecting a robot that has worked with toxic chemicals

Military robots help save lives, but they can also do harm. Some robots are armed with guns, missiles, and other weapons.

For now, humans control these robots remotely. However, people are figuring out how to create armed robots that are **autonomous**. That means they can find a target and decide to shoot on their own, without the guidance of a human.

Russian Platform-M combat robot

weaponized MUTT

THE DEVELOPMENT OF MILITARY ROBOTS

Some say Nikola Tesla created the first military robot in 1898, when he invented a radio-controlled boat. Unfortunately for Tesla, the military did not want to use his invention.

Nikola Tesla

radio-controlled boat

About 50 years later, during World War II, unmanned machines made another appearance. The **Soviet Union** had big Teletanks and Germany had small Goliath tracked mines, or "beetle tanks." Near the end of the war, the U.S. and Germany also both used unmanned bombers.

Goliath tracked mine

Soviet Teletank

German unmanned bomber

Hermes

The U.S. began launching military drones in the 1990s. The first use of a military robot in ground combat happened in 2002. The U.S. Army sent a small robot called Hermes into a cave near Qiqay, Afghanistan, to find out if it was safe.

Robots also helped the military in non-combat situations. After the attacks of September 11, 2001, small robots called PackBot and TALON helped search the rubble in New York City for survivors.

search and rescue after September 11 attacks

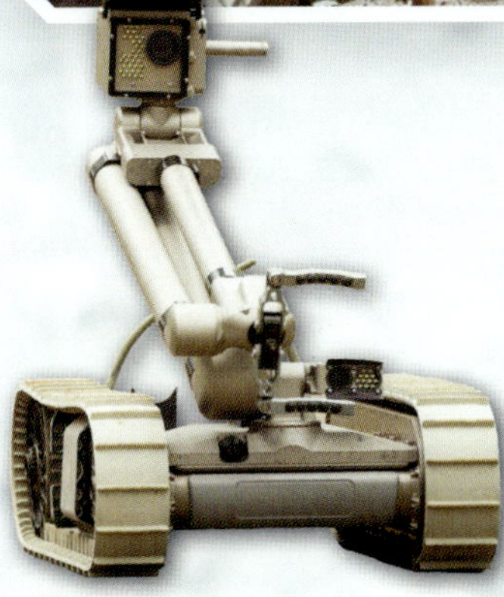

PackBot

TALON

Since 2002, military robots have become much more common. Computers have become smaller and faster, and **sensors** and other hardware have become cheaper. This makes it possible to make military robots more advanced and powerful.

People disagree about the safety of making more powerful robots. If robots have **artificial intelligence** (AI), they can make decisions on their own. Many think this will make robots more effective fighters. Others think they will be dangerous "killer robots."

KILLER ROBOTS?

A group called the Campaign to Stop Killer Robots hopes to stop militaries from using autonomous armed robots. Stephen Hawking and other famous scientists support this group.

MILITARY ROBOT PROFILE: PACKBOT

PackBot weighs about 40 pounds (18 kilograms) and fits onto a soldier's backpack. A soldier can unpack the remote-controlled robot and set it up in minutes. One of the most useful things about PackBot is that it can move over rough ground. It can climb over rocks and even up stairs. If it falls over, it uses side flippers to right itself and keep going.

Three versions of PackBot come with different tools for different jobs. The basic version is the Scout. Explorer PackBot has several cameras. The EOD's extendable arm finds hidden bombs.

ROBOFACTOID

Water will not hurt PackBot. If it drops into 3 feet (1 meter) of water, it will work fine when it comes out.

Name:	PackBot
Developer:	iRobot
Release Date:	2001
Functions:	gathering information, mine clearing, search and rescue
Size:	7 inches (18 centimeters) high; 27 inches (69 centimeters) long; 20 inches (51 centimeters) wide
Speed:	5.8 miles (9.3 kilometers) per hour
Reach:	6 feet (1.8 meters)

MILITARY ROBOT PROFILE:

LS3

The Legged Squad Support System (LS3) moves like a big dog. Its legs work on rough **terrain** and up and down hills. Marc Raibert and his team created it to carry supplies for soldiers in places where vehicles cannot go.

The LS3 is well trained. It does not need a remote control. It follows its human leader using its computer vision. The robot can also follow some voice commands or go to a certain location using **GPS**.

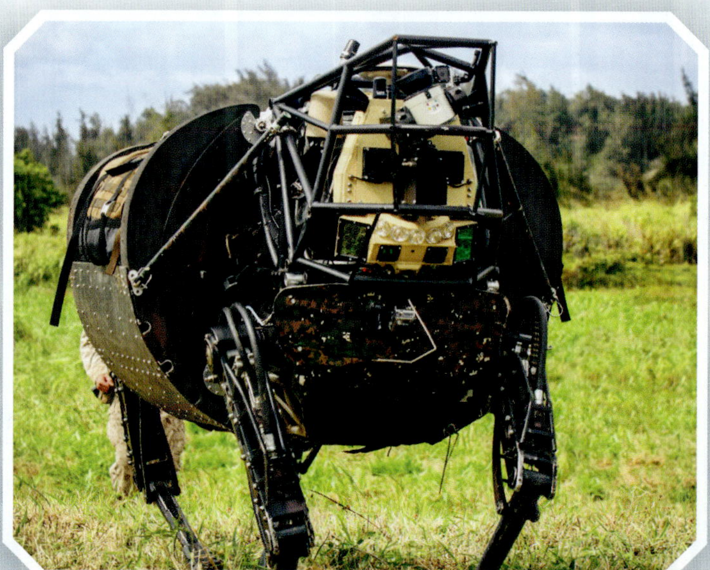

Name:	Legged Squad Support System
Nicknames:	LS3, AlphaDog
Developer:	Marc Raibert, Boston Dynamics
Release Date:	began field testing in 2012
Function:	carrying supplies
Size:	about 8 feet (2.4 meters) long
Speed:	7 miles (11.3 kilometers) per hour
Load:	can carry 400 pounds (181 kilograms)

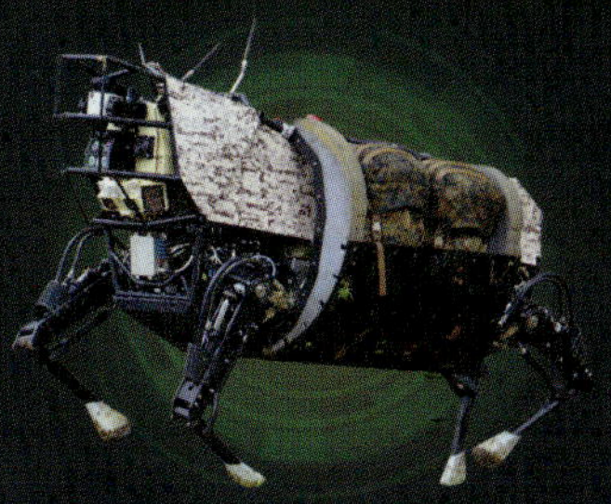

TURN IT DOWN!

After testing LS3, the military decided not to use it in combat. The reason? Its engine is loud. It would give away the position of the soldiers!

MILITARY ROBOT PROFILE: ROBART III

Unlike many military robots, the ROBART III is not used in battle. It was made to be used in the lab. Scientists use the **humanoid** ROBART III to figure out problems.

One problem ROBART III helped with is running many robot functions at once. ROBART III is a security robot. It can look for and detect **intruders**, aim a weapon, and drive. To control all these functions, researchers learned to connect them together. ROBART III also controls a team of smaller bug-like bots.

EARLIER VERSIONS

ROBART III has two older siblings. ROBART I only detected intruders. ROBART II detected the intruder and decided if it was a threat.

Name:	ROBART III
Developer:	SPAWAR (U.S. Navy)
Release Date:	1992
Functions:	research, security
Size:	about 4 feet (1.2 meters) tall

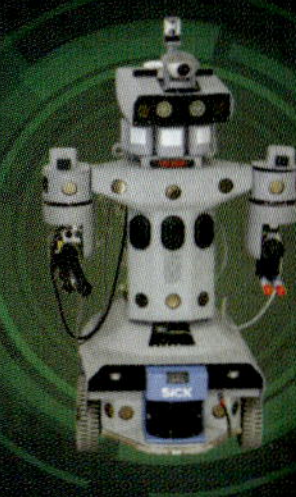

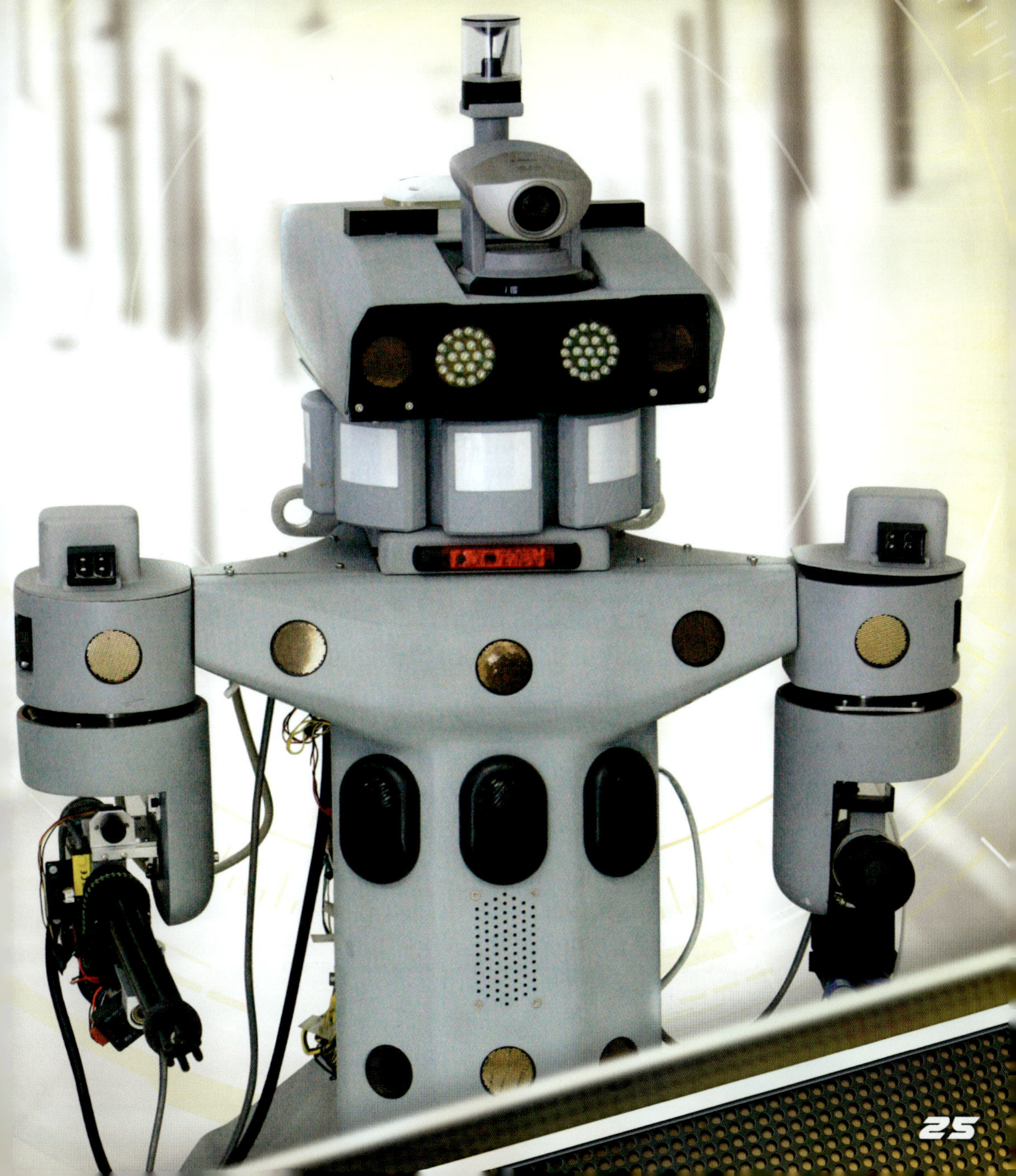

MILITARY ROBOT PROFILE:

TALON

TALON is another popular man-portable robot. Soldiers use it in dangerous situations. TALON can find bombs or search a cave or abandoned building. The robot is tough and sturdy. It can move through water and snow. It also climbs stairs and rock piles.

Some TALON robots come with upgrades. An upgraded TALON called SWORDS can carry **grenade launchers** and other weapons. The **HAZMAT** TALON has a special job. It uses special sensors to detect radiation and poisonous chemicals and gases.

A POPULAR ROBOT

QinetiQ has sold around 4,000 TALONs to countries all over the world.

Name:	TALON
Developer:	Foster-Miller (now part of QinetiQ)
Release Date:	2000
Functions:	bomb disposal, HAZMAT, search and rescue
Size:	34.5 inches (88 centimeters) long; 22.5 inches (58 centimeters) wide; 18 inches (45.7 centimeters) tall
Speed:	6 miles (9.6 kilometers) per hour
Reach:	77 inches (2 meters) vertical; 62 inches (1.5 meters) horizontal

THE FUTURE OF MILITARY ROBOTS

Military robots have saved many lives. Soldiers rely on them for help in the most dangerous situations. Because they are so useful, there are now thousands of them. About one-third of U.S. military aircraft are UAVs. More military robots are used each year.

New types of robots are on the way, including those with AI. Many people believe that robots that can fire without human command should be banned. Others are working hard to build such robots. The future of military robots depends on these decisions.

UNDERWATER ROBOT

The U.S. Department of Defense is testing a robotic ship called Sea Hunter. It will run without a crew for up to three months, searching for enemy submarines. It is expected to be part of the U.S. Navy in 2018.

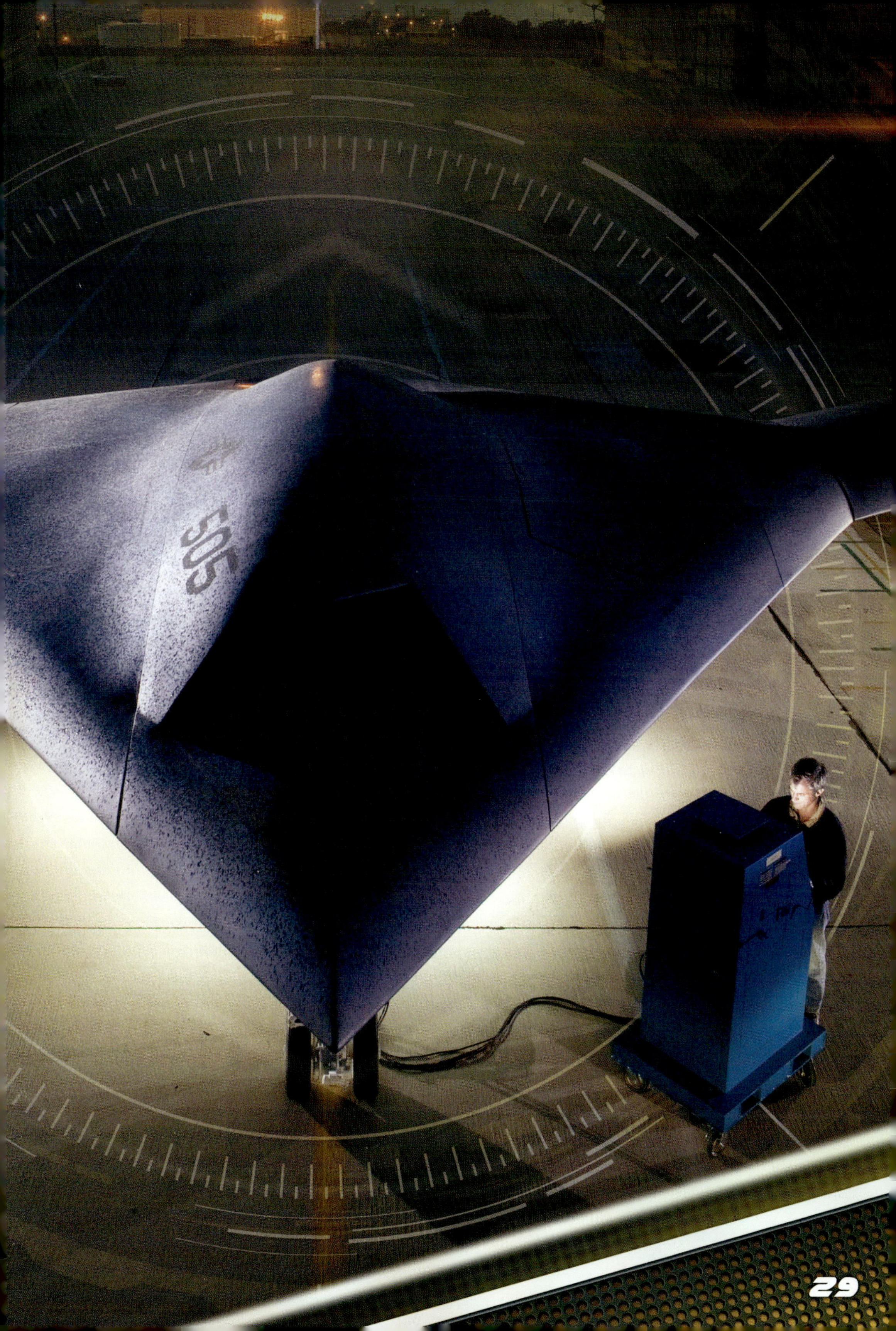

GLOSSARY

artificial intelligence—a computer's ability to do things a human mind can do

autonomous—able to act without help

base camp—a temporary home for soldiers

GPS—short for global positioning system; GPS is a navigational system that uses satellites to determine locations on Earth.

grenade launchers—weapons that fire small bombs called grenades

HAZMAT—short for hazardous materials; hazardous materials are substances that could hurt humans and animals or make the land or water toxic.

humanoid—having a human-like shape

intruders—people who are somewhere they should not be

land mines—bombs hidden in the ground that are set off by footsteps or other weight

missile—an object used as a weapon that is launched to hit something far away

nuclear radiation—harmful energy sent out by nuclear waste

sensors—devices that respond to light, pressure, sound, or other physical changes

Soviet Union—short for the Union of Soviet Socialist Republics, a former country in Eastern Europe made up of 15 republics and states

surveillance—the action of keeping a close watch on someone or something

terrain—another word for land

TO LEARN MORE

AT THE LIBRARY

Faust, Daniel R. *Military and Police Robots.* New York, N.Y.: PowerKids Press, 2016.

Mooney, Carla. *Awesome Military Robots.* Minneapolis, Minn.: Core Library, 2015.

Spilsbury, Richard and Louise. *Robots in the Military.* New York, N.Y.: Gareth Stevens Publishing, 2016.

ON THE WEB

Learning more about military robots is as easy as 1, 2, 3.

1. Go to www.factsurfer.com.
2. Enter "military robots" into the search box.
3. Click the "Surf" button and you will see a list of related web sites.

With factsurfer.com, finding more information is just a click away.

INDEX

The images in this book are reproduced through the courtesy of: Department of Defense, front cover, pp. 4, 4-5, 6 (top left and right), 6-7, 7, 8-9, 9, 10-11, 11, 12-13, 17 (bottom left and right), 18, 20, 20-21, 21 (top), 22, 22-23, 23 (top), 26-27, 27, 27 (top), 28; PA Images/ Alamy Stock Photo, p. 8; Goga Shutter, p. 12; Public Domain, pp. 14 (top and bottom), 14-15, 15 (left and right); Associated Press, pp. 16-17; New York Daily News Archive/ Contributor/ Getty Images, p. 17 (top); Oli Scarff/ Staff/ Getty Images, pp. 18-19; ZUMA Press, Inc./ Alamy Stock Photo, pp. 24-25, 25 (top); John B. Carnett/ Contributor/ Getty Images, pp. 28-29.